# The Earth and Relativity

by

John D. Merrifield

DORRANCE & COMPANY
Philadelphia

ISBN: 0-8059-1644-X
Library of Congress Catalog Card Number: 79-180937
Printed in the United States of America

## CONTENTS

## PREFACE

The object of this book is to formulate a logical answer to the basic question, "What is the relation of the earth to all other parts of the universe?"

If this book is successful in helping to bring relativity and earth science one step closer, each to the other, it will have served its purpose.

# Chapter I

## VELOCITY

Velocity more than any other single characteristic of nature determines the existence of the earth, the moon, the stars, the planets, the galaxies, the life of man, all living things and all things conceivable.

If we go for a trip in the mountains and see a large boulder of granite lying on the surface of the earth, we normally think of this boulder as being at rest. This boulder is not at rest, not even in relation to the earth. The mountain on which the boulder is resting is always on the move. It is rising or falling in relation to the center of the earth. It is moving closer to or further from the equator of the earth, and it is moving closer to or further from any reference point on the surface of the earth, such as Rocky Ford, Colorado.

Granite is a crystalline mixture of feldspar, quartz, mica and minor accessory minerals. The crystalline formation and texture of this granite boulder was determined, in a roundabout way, by the rotation of the earth. The moon does not rotate on its axis: therefore, it is logical to conclude that large blocks of granite with a crystalline texture similar to that which was formed in the earth will not be found on the moon.

It has been less than three hundred and forty years since Galileo was summoned before the Inquisition at Rome and forced to recant his belief that the earth moves about the sun. It has been approximately three hundred years since Newton was formulating his gravitational theories and it has been over sixty years since the Michelson and Morley experiments did much to confirm Einstein's theory of relativity.

The fundamental difference between the classical (Newtonian) theories and the contemporary (Einsteinian) theories is that in the days of Newton it was demanded by the physicists that any theory, to be accepted, must be such that a mental picture could be drawn of the phenomenon in question or that a mechanical model could be constructed; whereas, Einstein's mathematical formulas deal with something which does not occupy space, cannot be seen or heard, and cannot be detected by its odor. Under such conditions it is impossible to use objects which occupy space to construct a mechanical model or construct a visual picture of this force of unknown construction. However, the effect of this force upon objects as they travel in the fields of the universe can be studied and mathematically explained.

The theory as to how the world is made is not essentially different today than it was in the days of Newton. Classical physics considered that all material objects were a formation of specks of an unknown construction. Contemporary physics considers that all material objects are a formation of some kind of a thing of unknown construction. To this thing of unknown construction the name *energy* has been given. In classical physics, the tiniest of specks were to be considered as material objects which occupy space and were not to be considered as interchangeable with any thing which did not occupy space. In contemporary physics, this energy of unknown construction of which we are made does not occupy space until it has become orientated into the tiny specks of classical physics. When properly orientated, this energy will assume the properties of mass. It will occupy space and be visible. In this condition it can be used to construct mechanical models or formulate mental pictures.

Since the days of Newton there has been a vast quantity of knowledge gained about nature and nature's ways, more especially in the field of velocity and its effect upon animate and inanimate bodies.

Contemporary physics has added much to classical physics but it is not much closer to the end of gainable knowledge than it was in the days of Newton. To bring the mathematics of the Newtonian laws of physics into compliance with contemporary laws of physics it will require many additions, such as, the length of the yardstick varies as its velocity in space changes. The weight and mass of a body vary as its velocity in space changes. Time is not a constant. What was thought to be a straight line is now considered to be a curved line. The expression "Straight as an arrow" cannot be used in contemporary physics because the shaft of an arrow, by its own free will and accord, will curve to match the curvature of the space energy field. If the arrow is rotated about the axis of its shaft the curvature of the shaft will not rotate with the arrow. The curvature of the shaft will remain stationary and conform to the curvature of the space energy field. The curvature of space is not a constant. When the arrow is shot into the air the curvature of the shaft will automatically change to match the curvature of space at each instant that it is traveling in space. To give an illustration, consider that an arrow is to travel close to a large inertial mass, such as the sun; the shaft of the arrow will automatically change its curvature as it approaches the sun. When the arrow is directly opposite the sun the change in its curvature will be at the maximum, and the change in the curvature of the shaft will become less as it recedes from the influence of the sun's highly concentrated inertial mass field.

During the past sixty years there has been a gradual increase in the rate of the acceptance and the development of the Theory of Relativity. Perhaps within the next sixty years there will be developed a much broader understanding and a greater acceptance of this theory and its relation to earth science.

## Chapter II

## RELATIVE VELOCITIES

In mechanics, velocity is defined as a rate of motion in which direction as well as speed is considered.

Man-made velocity in relation to the earth in which we live is well understood and fully comprehended. Velocity within the earth relative to the total universe is the most important but the least understood of all the branches of modern earth science.

In the space energy field, the velocity of a body of matter relative to the velocities of all other bodies of matter will determine the amount of space occupied by the said body of matter. The length, breadth and thickness of all bodies of matter are determined by their velocities through space in the direction at which their velocities are under consideration.

Albert Einstein devised the following formula:

$$L = L_1 \sqrt{1 - \frac{v^2}{c^2}}$$

In this formula L equals the length that any body of matter will have as it moves through space. $L_1$ equals the length the said body of matter would have, should its velocity be zero in relation to all other bodies moving in the space energy field. V equals the velocity of the body of matter which is under consideration as it travels in the space energy field. C equals the velocity of light.

In the studies of energy it has been proven that energy is discontinuous and does not flow in a continuous stream. All energies flow in a quantum of energy. Since energy flows in quantum, Einstein's formula for calculating length must require that the unit of measurement for length be in quantum. Any length less than one quantum would be zero. In the above formula when the velocity of a three-dimensional body becomes equal to the velocity of light, its length will be zero because the square root of one minus one is zero. This zero does not mean that absolute zero has been obtained. It means that the distance between particles is less than one quantum of length. All that is left could be classed as free energy, i.e., energy which does not occupy space. If Einstein's theory, that mass and energy are interchangeable quantities, is correct the above could be a condition where mass has turned into energy. In other words mass and energy are interchangeable quantities at this condition. The universal field is operating inside the earth with the same characteristics that it is operating in the space outside the earth. It is therefore logical to believe that it is possible for energy and mass to become interchangeable quantities inside the earth. The astronomers say that this action is continually taking place in the heavens.

Our world is rotating and it is traveling in various directions through the space energy field. This formula says that every dimension of any molecule of matter is changed each time there is a change in the velocity of the molecule when this velocity is measured in relation to the space energy field. The field of the inertial mass energy cannot be considered as stationary because it is affected by each and every body of matter and all of these bodies of matter are traveling in various directions in the universal field. For this reason no mathematical formula can be written which will use as its base of operation some imaginary stationary point in the universe. This should not stop the coordination of scientific reasoning from interpreting the full meaning of Einstein's master formulas.

As an illustration of velocities consider the wheel of an automobile traveling along the road at 60 miles per hour. The velocity of the axle of the wheel in relation to the earth will be 60 miles per hour. Its velocity in relation to the moon will not be 60 miles per hour because the earth is rotating and the moon is circling the earth. Its velocity in relation to the sun will not be 60 miles per hour because the earth is rotating and the earth is circling the sun. Its velocity in relation to the inertial mass energy field will not be 60 miles per hour for various reasons, such as that the earth is rotating and circling the sun, the sun is traveling through space, the inertial mass field is expanding, and all bodies in space are on the move.

The velocity at the bottom of the wheel in relation to the earth will be zero. It is zero because the wheel is not slipping on the earth. The velocity of the top of the wheel in relation to the earth will be two times the velocity of the axle of the wheel. The velocity of the forwardmost point of the wheel in relation to the earth will be the square root of the sum of 60 squared plus 60 squared. The velocity of the rearmost point of the wheel in relation to the earth will be the same as the forwardmost point of the wheel but its direction of travel will be different. (This direction of travel is a fundamental and very important basic characteristic.) All of the points on the spinning wheel will have a distinct velocity, relative to the earth. This relationship between the spinning wheel and the earth will be same today as it was yesterday and as it will be tomorrow. This does not remain true for the inertial mass field because the inertial mass field is always changing its direction and its magnitude of travel in relation to any conceivable point of reference.

The mathematical relation between the inertial mass field and a wheel rolling on the earth cannot be determined because the inertial mass field is not a constant nor has it been mathematically captured. For this reason, Einstein's ultra-simple formula for determining relative lengths in relation to relative velocities becomes of the utmost importance for the study of geology and earth science.

It is the relative velocities of the molecules of the earth in relation to the relative velocities of all other molecules of the universe which determine the forces which create the structure and the conditions of the crust of the earth. It just happens that these structures and conditions are such that man in his present form can survive.

The illustration of the wheel rolling on the surface of the earth which was used to explain relative velocities is so simple that it might be considered by many people as very elementary. It is however, very important that this fundamental principle of a spinning wheel and its relation to earth science be given serious consideration.

If the wheel rolling on the surface of the earth used in the previous illustration to show relative velocities were placed at the center of the earth and it continued to spin, each molecule of the wheel would have a constant resultant velocity relative to all other molecules of the earth. The rotation of the earth would not be increasing or decreasing the magnitude of the resultant velocities relative to the spinning wheel.

Likewise should our earth be the center of the universe there would be no appreciable amount of lengthening and shortening of the molecules which constitute the earth. The relative velocities of the spinning earth in relation to the space energy field would be similar to the velocities of the spinning wheel, in relation to the earth, i.e., when the wheel was located at the center of the earth. Therefore, there would not be a continuous expansion and contraction movement beneath the crust of the earth to produce the friction necessary to generate heat. As a result of the absence of generated heat, the crust of the earth would become cold as compared to its present temperature. This shows that our earth is not at the center of the universe and perhaps it is not close to the center of the universe. There are indications, as shown in my first book (*Mountains, Valleys, and Subterrestrial Forces**), that the center of the universe, for the present one-billion-year cycle, has been moving somewhere in a plane which might be intersecting the earth at an averaging angle of approximately 65 degrees from the north relative to the earth's axis.

The heat generated by the stretching and the compressing of the molecules of the earth has no relationship with entropy or the degradation of energy. It is a part of the steady state of all of the energies of the total universe.

The total energy of a three-dimensional body of matter is, according to Einstein, equal to its mass times the velocity of light squared. Heat is a part of the rearranging of the energy in the molecules and it will comply with Einstein's formula for mass and energy.

* Philadelphia: Dorrance & Co., 1970.

The calculations for determining the velocity of inertial mass particles, relative to all the other bodies of inertial mass floating in the universal energy field are so complicated that no mathematician can devise a mathematical formula to explain the necessary steps which must be taken to prove or disprove the theory of relativity. The inertial mass field in relation to the space energy field is continually expanding in all directions. Thus, billions upon billions of inertial mass particles are moving in many unchartable directions.

The inertial mass of a particle of matter is determined by its relative velocity in the space energy field. Since the relative velocity of matter cannot be determined, its inertial mass must be an indeterminate. The theory of relativity does, however, explain how velocity determines the mass of a particle of matter. It considers the mass of a particle of matter equal to its energy divided by the velocity of light squared.

There are many happenings inside our earth which cannot be explained without using the theory of relative motions. The most important is the creeping magma. The creeping magma forces mountains up, pulls trenches down and furnishes fuel for volcanoes. It moves some sections of the surface of the earth toward the equator and other sections away from the equator. And it increases the distance between some of the continents.

As an illustration of relative length, let us examine the length of the yard. By definition the length of the yard in the United States is derived from the prototype meter kept in the National Bureau of Standards and it is related to the meter, which is kept in Paris, by the ratio of 3600 to 3937. When the yard and the meter are placed side by side the above ratio is correct but when the meter is in Paris and the yard is in Washington the above ratio does not hold true for all hours of the day. The world is rotating and traveling through space. There are, therefore, times when the velocity in space at Paris is greater than the velocity in space at Washington and there are times when the velocity in space is less at Paris than it is at Washington.

It has been proven by the Michelson and Morley experiments and other measurements that the length of the prototype meter after it was moved from Paris to Washington does not conform to the ratio of 3600 to 3937. However, the discrepancy is so slight that it cannot be observed unless the measuring stick is exceptionally long, such as the magma layer which is circling the earth. Here the creep of the magma layer is the indicator that various sections of the magma do not remain a constant length.

All of the velocities observable in nature are taking place in curved lines; therefore, it is impossible to construct mathematical formulas based on straight lines and transformations to calculate the difference in length of the meter stick located in Paris and its prototype located in Washington. Likewise, the same discrepancies hold true when trying to compare the weight of an object located in Paris to an object of equal weight located in Washington.

## Chapter III

## MAN-MADE VELOCITIES

In the second chapter velocities relative to the velocities of the universe were studied. In this chapter velocities as changed by man will be examined.

Velocity determines temperature, mass, chemical characteristics, mechanical characteristics, electrical characteristics, the specific kind of an element of matter, gravitational attraction, inertia, length, breadth and thickness of all materials which occupy space and all the properties of matter.

In the chemical laboratory, the chemist can mix hydrogen and oxygen to obtain water. The chemist can mix many elements to obtain chemical compounds. He can influence or change many chemical actions by changing the environment. The chemist cannot manufacture gold by the mixing of chemical elements, because the mixing of chemical compounds does not change the velocity of the building blocks of the elements of matter which occupy space.

The alchemists for thousands of years spent much time trying to manufacture gold by mixing chemical formulas. They were always in the hope of finding the elixir for changing a base metal into gold. They performed chemical experiments in every conceivable manner which were possible under known conditions. They could see that nature was changing the arrangement of the building blocks of matter and that this new arrangement of the building blocks was producing gold, copper and many materials which occupy space. They never gave up trying to obtain transmutation of the elements of matter. They were never close to success. The reason that they did not obtain the slightest degree of success was that they did not understand that the correctness of relative velocities, and not environment, is the condition which will synchronize the energies and hold them properly arranged for nature to build the elements of matter.

At the turn of this century it became evident to all scientists that velocity instead of environment has the most control over the changing world. They learned that to manufacture gold, it will be necessary to use the science of mechanics and exert an energy force upon the building blocks of matter and thereby change their velocity. When all the internal velocities of the various building blocks are correct for their combining, it becomes possible for nature to manufacture atoms of matter such as gold, copper, tin, et cetera.

The process is not as simple as one might assume from the foregoing explanation. Let us assume that it could be possible for man with the aid of a pair of tweezers to pick the proper specks from one element of material and throw these specks, at the proper speed, into another element of material. This element would then be changed into the element named gold. If a speck of material is thrown at a too-high velocity, it will pass through the element of material without stopping. If the speck of material is thrown at a too-low velocity, it will bounce off the surface and not enter into the atom of material which is being changed to gold.

It is not possible to accomplish the above with tweezers. It is, however, possible to use electrical energy to increase the velocity of an element of material until it is easily taken apart by mechanical means and then to slow it down by bouncing it from a deflector of the proper resilience and thus manufacture gold. This is a strict mechanical operation and has no relation to chemical reactions.

The reason that gold is not now manufactured for commercial use is that it will cost approximately nineteen times as much to manufacture gold as it does to dig it out of the ground and refine it.

For thousands of years previous to this century heat and pressure were the two principal laboratory tools used in the study of the structure of the atom and the molecule. The temperatures and the pressures obtainable in the laboratories were so small as compared to the temperatures and pressures existing in the earth that no indication of a similarity could be detected between the actions taking place in the earth and the actions taking place in the laboratories.

At the turn of this century the machinery for producing electricity had been developed to such a degree that electricity became the major energy for studying the velocities of the atom. It became the principal tool for controlling the velocities within the atom. It has simulated many of nature's conversions of the energies. As an illustration, man manufactures gold in the laboratory, nature manufactures gold in the earth and the astronomers observe that this kind of manufacturing process is taking place in outer space.

When man increases the velocity of an atom until it approaches the velocity of light and then dissects it, there is a quantity of inertial mass which has interchanged with energy. This newly formed energy escapes from man's control and becomes a part of the energy which constitutes the universal energy field. In the universal energy field it does not possess gravitational attraction or a velocity of its own. It accepts and conforms to the velocity of the universal energy field. It relinquished all ties with inertial mass and floats along with the other nonseeable energies. It does, however, have the ability to make its appearance as inertial mass, should there be an opening for it, at any point in the universe. As an illustration, it could become inertial mass should it happen to be at the right place at the right time in a field where nature is creating inertial mass from the energy of the space energy field. This kind of a manufacturing process is understandable. Most astronomers contend that hydrogen atoms are being produced, in large quantities, in the galaxies of the universe.

## Chapter IV

## THE QUANTUM THEORY

The quantum theory of energy is a theory that energy is not a smoothly free flowing continuum, that it has a discontinuity, that energy is radiated or absorbed in bundles, and that the mathematical amount of energy in each bundle is always equal to a constant.

The mathematical constant referred to is known as Planck's constant, which is .000,000,000,000,000,000,000, 000,006,62 ergs per second.

This constant was arrived at by experimenting with the degradation of energy on the surface of the earth. It is an extremely small constant but if a constant could be determined for the space energy field at the instant when energy is converting to mass, or when mass is converting to energy, it is likely that this constant would be many times smaller than Planck's constant. This action would probably be almost instantaneous. The action would be completed in one quantum of time. It is likely that this constant will never arrive at absolute zero because if the constant were zero, energy would become an uncontrollable fluid and would not possess any specific structural arrangements. Consequently, the earth and the other planets would cease to maintain a structure which occupies space in the universal field.

It was not necessary to know the mechanical structure of energy when determining planck's constant. Planck's constant will apply equally well to the pellet theory of energy, the wave theory of energy, the pinwheel theory of energy, or the pulsating theory of energy. There is nothing in any one of the four theories of energy that can prove itself to be more explicit about energy than the other three theories; therefore, one can assume that none of the four theories of energy is fully explicit in all details. The planetary theory of the atom, commonly known as the Bohr Atom, has done much as the guiding idea of the quantum theory. It has worked well for the study of the energies of the atom. It has, however, not been very helpful in the study of the surfaces and angles of crystals. Crystal surfaces with the perfection of smoothness and sharp angles do not build very readily with planetary motions. The book titled *Crystals Perfect and Imperfect* by Westinghouse Research explains an action which could be likened to atoms connected by coiled springs and explains that this arrangement could produce the orderliness in crystals.

Even though a mechanical explanation of energy may not be fundamentally correct in detail it is necessary to create some kind of a mental picture for one man to convey his thoughts to his fellow man. The object of a theory of energy is to calculate results that can be compared with experiment, and it is not necessary that any satisfying description of the whole course of energy phenomena be known or be stated. Planck's constant supplies the mathematical tool for quantitative reasoning.

From the experiments conducted in the laboratories it is concluded that energy must be transported in bundles and that each bundle has no connection to the previously delivered bundle. It does, however, bring changes to the total energy field. The most important tools in use for studying the effect that the energies from outer space have upon our earth, are the energy spectrum, mathematics, and the effect that the changes in relative velocities have upon the atoms of materials which compose the elements of the earth.

Light energy is a portion of the energy spectrum; therefore, some comments about light energy are in order. Light energy has length and thickness, but its width is less than one quantum of discontinuity; therefore, it can be rightly called two-dimensional. As an illustration, if two parts of light energy are taken from a single ray of light, with one part coming from the positive side of the ray and the other part coming from the negative side of the ray, and these are passed through two holes in a screen, these two parts can be made to neutralize after they have passed through the two holes.

In the above experiment, if a single ray of light is used to obtain scintillation, it will be logical to conclude that the two holes in the screen must lie in the same plane as the ray of light from which the plus and minus parts of the light energy ran were taken. Should the line intersecting the two holes in the screen be at right angles to the plane from which the light ray were taken. Should the line intersecting the two holes in the screen be at right angles to the plane from which the light ray was taken it is concluded that it would be impossible to obtain scintillation. This would show that a portion of positive light cannot be neutralized by a portion of negative light when their light planes are at right angles to each other.

Mathematics is the most far-reaching and important of all the tools used in studying the energies of the universe. By the use of mathematics, it can be shown that a certain thingamabob must exist or that some kind of an action or reaction is taking place. After the researcher knows what he is looking for he goes out and finds it.

Mathematics was used to show that the velocity of a three-dimensional material body traveling through space would determine the length, the breadth, and the thickness of that body. The Michelson-Morley Experiment was one of the many different kinds of experiments performed which substantiated this part of the theory of relativity.

According to the Theory of Relativity, the velocity of a body of matter relative to all the other bodies of matter traveling in space will determine the quantity of space occupied by said body of matter, such as a steel beam. It can, therefore, be concluded that the energy of space is at least a three-dimensional energy.

In keeping with the theory of relativity it is permissible to say that a quantum of a two-dimensional energy which does not occupy space in the space energy field can become a quantum of a three-dimensional energy which will occupy space in the space energy field. This is merely one more way of stating Einstein's Theory that "mass and energy are interchangeable quantities."

Planck's constant mentioned at the start of this chapter has discrepancies when applied under extreme conditions.

This constant was arrived at by experimenting with two-dimensional energies. The principal energies used were in the field of specific heat and photoelectricity. Because of the numerous inexplicable exceptions to Planck's theory the theory of quantum mechanics was developed. This theory is essentially mathematical in nature and has fewer arbitrary assumptions than Planck's quantum theory. Quantum mechanics, which explains intensities, is now having more success than the quantum theory is having in the explanation of the structure of the atom. All of the energies which have been used in developing both the quantum theory and the theory of quantum mechanics were of two dimensions in nature. Two-dimensional energies were used to pull three-dimensional energies apart and two-dimensional energies were used to force these energies back together to form a new and different three-dimensional mechanical structure.

It is logical to believe that when a two-dimensional energy is interchanging with a three-dimensional energy to form a three-dimensional body of mass, neither Planck's constant nor the quantum mechanics theory can be fully explicit.

At the instant of an energy interchange, the elapsed time to be used in calculating quantum mechanics will be zero. This is true because any elapsed time less than one quantum of time is, in quantum mechanics, zero time. Also any time less than one quantum of time can be classed as instantaneous. This shows that there can be an instantaneous action at a distance with no interconnecting medium. Energy not bundled can travel with no interconnecting medium. It therefore complies with Einstein's Theory of Relativity.

When energy which does not occupy space and does not possess gravitational attraction becomes properly bundled it will occupy space and possess gravitational attraction. The quantity of space now occupied by the mass will be determined by its velocity in space relative to all the other bodies of bundled energy traveling in the space energy field. Likewise its inertia and gravitational attraction will be determined by its velocity in the space energy field relative to all other bodies traveling in space.

When nature is building the hydrogen atom in the galaxies,

as the astronomers contend that it is, Planck's constant will not apply at the instant when free energy starts on its trip to become bundled energy. Man cannot detect any movement which takes place in a quantity of time which is less than one quantum. At the instant when a quantity of energy is leaving the field of the free energy to enter into the field of the bundled energy to become inertial mass its velocity relative to the two fields is zero. As an illustration, in a train of gears in which the teeth of a small gear are meshed with the teeth of a large gear the relative velocity at the point of the touching tangents of the two gears in relation to the seeable energy field is the same. In other words the velocity relative to each other is zero. As soon as the point of touching tangents is broken, the relative velocity is no longer zero. When the relative velocity was zero Planck's constant would not apply because they both had the same velocity relative to the inertial mass field. In one quantum of time after the point of touching tangents is broken, Planck's constant will apply because this newly created inertial mass has now become a part of the seeable mass and produces its share of the inertial mass field of matter. This action will be put to use in the chapter under the heading of Crystallization.

Very little has been included about quantum mechanics in this dissertation. A quantitative theory of quantum mechanics has not been developed to apply to quantums which have a value less than Planck's constant. The whole nature and orderliness of the seeable universe is dependent upon a constant of energy which has a value as large as Planck's constant. Any packet of energy less than Planck's constant will not affect any of our systems of detecting energy. Our systems of detecting and measuring quantities of energy are based upon the theory of atomic radiation. Any packet of energy which has a value of anything less than Planck's constant is not a part of the seeable universe. It can, however, be a part of the free and unseeable energy floating throughout the seeable universe and beyond into the nonseeable universal energy field.

Early in the twentieth century the equipment for energy experimentation had been developed to such a high degree that it enabled man to satisfy his thinking that light energy and all other forms of energy move in bundles. To explain this phenomenon, the quantum theory of energy was devised. The value of the smallest bundle of energy which will affect our present day machines for measuring energy is known as Planck's constant.

This constant applies to all the energies of that part of the universe which is seeable or measurable with our present day equipment. To give a quantitative meaning to all atomic energies, the theory of quantum mechanics has been highly developed and put to much use. Quantum mechanics shows that energy with a quantity value which is less than Planck's constant leaves the measuring machine and escapes from man's control when the atom is disintegrated. For this transfer of energy the quantum theory that energy is discontinuous will still apply but the theory of quantitative mechanics based on Planck's constant will not suffice.

In the inner earth there is a mechanical oscillating force at work. This mechanical force which can be likened to a tractor pushing a snow plow is pumping away at the packets of energy. When there is a void in the energy of the seeable universe this pumping action forces a quantity of energy from the nonseeable energy field into the seeable energy void. This energy transfer of nature has no relation to the principle of the degradation of energy. It does, however, have a distinct bearing upon the steady state of the universe.

Most scientists agree that should Planck's constant become zero there would be no stability and organization of the universe. When studying the real meaning of this statement, one must take into account that the quantum theory is based upon the wave of motion theory and that all wave theories recognize that there is an instant of time at which the kinetic energy of a wave is zero and its potential energy is at the maximum. The zero referred to is the instant when the mass particle is at the end of its stroke, i.e., the maximum distance from the center line of a wave motion. It is at this instant of time in which the steady state energy of the universe is influenced to leave the unseeable energy field and enter into the seeable energy field of the universe. All that will be required to start this energy transfer will be a slight inequality of mechanical equilibrium inside the structure of the inertial mass particle. The force multiplied by the distance traveled will be less than Planck's constant. However, by the end of the first completed quantum of time the potential energy of the space energy field will have been converted to kinetic energy in the seeable energy field; therefore, the total energy transfer will have become equal to Planck's constant. This energy change can take place with little persuasion because the velocity gradient between the two fields at this point is zero. In other words they are at rest in relation one to the other at the instant when inertia receives its meaning. The transfer of energy from the nonseeable energy field into the seeable energy field to become inertial mass is a mechanical operation. When this kind of energy transfer takes place in the earth it must be the result of a mechanical stress and strain relation within the structure of the two forms of energy.

## Chapter V

## ENERGY

There are several definitions in the dictionary for the word *energy*. In the study of physics the word has generally meant "the capacity of doing work and overcoming resistance." This definition is a very good one for the study of the degradation of energy.

This dissertation on the study of energy is in relation to the steady state of the universe: therefore, when one speaks of energy in relation to the universal energy of the space energy field where mass and energy are interchangeable quantities it becomes more meaningful to use the definition which states that "energy is the inherent power to produce an effect." The effect referred to would be the rearrangement of the energies of the steady state of the universe.

When we try to explain the energies of the universe it becomes very difficult to find the proper words because a language for steady state energies has not been developed. A word used to express an idea or existence of a state of energy might have one meaning to a man thinking in terms of the degradation of energy and an entirely different meaning to a man thinking in terms of the steady state energy of the total universe.

When we speak of energy in relation to Newtonian physics, i.e., physics before the turn into this century, it means that the energy of a moving body of matter is measured in relation to the earth. Mathematically speaking this would be one-half the weight of the body multiplied by its velocity squared, i.e., its velocity in relation to a point of reference on the earth.

When we speak of energy in relation to the space energy field it means that the velocity of the moving body must be measured relative to the velocity of all other bodies moving throughout the universe. Mathematically speaking, Albert

Einstein's formula for this condition signifies that the mass of a body can be determined by dividing the energy of the body by the velocity of light squared. This is true because the velocity of light is the same as the velocity of energy in the space energy field.

Energy, the inherent power of the universe, has never been isolated in the laboratories and experimented with. The effect of the inherent power of the universe has in many scientific ways been studied and the results noted.

During the latter part of the nineteenth century many scientists were trying to prove that electricity was the basic energy from which all other forms of energy and matter were derived. Close to the start of the twentieth century the electron was isolated and weighed. This achievement added confidence to many people who had believed that eventually it would be shown that electricity was the building block from which all matter could be created.

Evaluating our present knowledge about the energy of the space energy field it is logical to conclude that electricity is only a two-dimensional measurement of a disturbance in an energy which has at least three dimensions.

For the last one-half century the Einstein doctrine has been receiving many verifications from observable phenomena. Most of the recorded events which substantiate the theory of relativity have been taken from beyond the earth. Many of our laboratory experiments in the field of atomic physics are now being explained by the Theory of Relativity.

Einstein's formula for energy states that the total energy of a three-dimensional body of matter is the mass of that body multiplied by the velocity of light squared. The Einstein doctrine states that in the above formula mass and energy are interchangeable quantities.

By using the mechanics of relativity, the structure of the atom, the molecule, and the masses of the earth can be understandably explained. The Theory of Relativity states that "mass is nothing other than a form of appearance of energy."

From the above it is logical to conclude that energy can appear in a two-dimensional form, or in a three-dimensional form or it can disappear into a one-dimensional form. A one-dimensional energy cannot occupy space, it cannot possess gravitational attraction, and it cannot possess inertia. The velocity of a unidimensional energy can, therefore, exceed the velocity of light without violating the Theory of Relativity.

An energy with only one dimension can be one of the universal energies which is traveling throughout the space energy field. It can also be the energy required to furnish the binding energy required to tie energy into bundles of energy to produce atoms.

When man disintegrates the atom there is a quantity of energy which has not been captured. The sum of the masses of the component parts of the atom is not equal to the mass of the original atom. This energy which has not been captured and accounted for could be a unidimensional binding energy. This energy would not occupy space. It could have an instantaneous action at a distance with no interconnecting medium and it might be the energy which had forced the energies together to form mass, or if you are of the school who would rather call it the energy which holds two electrons of opposite potentials apart, it will be in keeping with the line of reasoning to call it that.

In cases where the electrons appear in pairs, and produce the pair fields, they appear in actual pairs of opposite potentials. They are maintained in pairs by the energy field but they do not make contact with each other. To construct a true mechanical model to illustrate this action is beyond the ability of the mechanical mind. Illustrations are however desirable. They will help the mind to follow a train of thought as it tries to understand the natural energy of the universe.

Many textbooks describe the pellet type theory of energy, the wave type theory of energy, the pinwheel type theory of energy and the pulsating type theory of energy. To these let us add the flutter type theory of energy. This type of energy would be a combination of the wave theory of energy and a portion of the pinwheel theory of energy. It could be said that the wave would trigger the instant for the flutter to take place and the flutter would instantaneously transfer the energy in the pair field.

In the pinwheel type of energy the two electrons with opposite charges are supposed to revolve around a center. In the flutter type of energy the two electrons with opposite charges would be pushed apart by a single-dimension energy which might have qualities likened to a coiled spring. They would not rotate. They would flutter in three dimensions. The transfer of energy within the pair field can be instantaneous without any connecting three-dimensional medium. The flow of energy from field to field cannot exceed the velocity of light but within a pair field the velocity can be infinite, i.e., any quantity of time less than one quantum of time can be classed as instantaneous. It is impossible to construct a three-dimensional model to explain a single-dimension energy which does not occupy space; therefore, none of the five above types of energy can be fully explicit.

The only thing in favor of using the flutter type theory of energy is that in the mathematics of quantum mechanics it will require a fewer number of assumptions than to use any one of the other four types of energy. In quantum mechanics there is no reason to prefer particles to waves or to prefer waves to particles. A combination of the two theories with a rigid connection between the waves and the particles might help to ease the mind when trying to understand the energy relation between a space energy field and an inertial mass energy field.

Regardless of the system used, when calculating quantum mechanics, if the answer obtained conflicts with the basic formula, i.e., "mass is equal to the total energy divided by the velocity of light squared," that answer will be in serious difficulty.

When calculating the degradation of energies in relation to this earth the formula that "energy is equal to one-half the mass times its velocity squared" is as true today as it was in the days of Newton. When the steady state energies of the universe are to be calculated it becomes an endless task to make the billions upon billions of calculations to be added to the degradation of energy formula to change this formula into a steady state energy formula.

The steady state energy formula that "mass is equal to the energy divided by the velocity of light squared" is so simple that it becomes an endless task to make the necessary calculations to prove the correctness of this formula. This formula is so simple when regarded from the point of view of relativity that it does not furnish a train of thought by which the human mind can follow to obtain a mathematical proof. It is, however, the answer and it furnishes the guiding principles for the mutation of elements within the earth. As an illustration, granite rock which consists chiefly of crystalline quartz, feldspar and mica is never found in the earth at a location above the magma layer which is resisting tension. It is always found above a location where the energies, which constitute the magma layer, are resisting compression.

It is not for this book to enter into the pro and con discussions regarding the theories leading up to the Theory of Relativity, mathematical or otherwise. It accepts the mathematical formula that "mass is equal to the total energy of the particle of matter divided by the velocity of light squared." It accepts this formula as the correct answer to a problem which cannot be fully explained by our present day language.

This book is primarily concerned with the actions taking place inside the earth. For this reason it will be in order to comment about the disagreement of scientists in regards to the value of pi (3.1416 +) as the ratio of the circumference to the diameter of a rotating disc.

As an illustration, consider there is a disc not rotating on the surface of the earth and that the circumference of this disc is a true circle. According to all mathematicians the ratio of the circumference to the radius would be pi and there would be no disagreement in this conclusion. As soon as the disc is rotating the circumference of the disc will discontinue to be a true circle. When the circle ceases to be a true circle the value of pi must be changed to conform to each point on the circumference of the disc. The change in the deformity of the circular circumference of the disc from rotating to non-rotating would be so slight that there is no known way of measuring this deformity. The amount of deformity of a rotating disc on the surface of the earth will be different than the amount of deformity of the same rotating disc on the surface of the moon. This is true because the velocity of the various points of the rotating disc in relation to the space energy field will be much different on the surface of the moon than it will be on the surface of the earth. This, perhaps, cannot be proven mathematically but there is much evidence, by the actions taking place within the earth, that this answer is a logical answer. After all is said and done it is the answer which is important and not the solving of billions of mathematical equations to arrive at a mathematical proof. When an action which is known to be occurring inside the earth cannot be explained by any known method other than the Theory of Relativity it is logical to use that theory until some better theory can be devised to give a more reasonable understanding of the known facts.

To mention some facts, it is known that mountains are being pushed up from the crust of the earth. It is known that some of these mountains are being pushed up faster than they are deteriorating. It is known that some of the trenches in the ocean are depressing faster than they are filling with sediment. It is known that there has been volcanic eruption for more than a billion years. It is known that continents are shifting on the surface of the earth, and it is known that large segments of the crust are twisting as they move. Relativity is the only single theory in existence which will explain all of these facts.

There are many pros and cons being presented by scientists about the velocity of light and how the light is propagated. There are two principal theories generally discussed. One is called the ether theory and the other is called the source theory. In the ether theory, commonly known as the luminiferous ether theory, the ether is supposed to be a hypothetical medium which is stationary in the cosmos and through which all light energy waves are transmitted. In the source theory the question is, does the center of an expanding light energy wave expand with the source of the light producing energy or does it remain stationary in the cosmos? Each of these two theories is confusing. They are in direct conflict, one with the other. Perhaps neither of the two theories is correct. Before any theory about the transmission of light can be fully explicit it will be necessary to know the construction of light energy and also the construction of the space energy field.

When man thinks about energy in any form such as light energy, heat energy, electric energy, magnetic energy, atomic energy, chemical energy, space energy, field energy, binding energy, et cetera, he cannot construct a hypothetical model to illustrate his thoughts because energy does not occupy space. He cannot see it, hear it, smell it, feel it, or know of its presence. He can, however, know of its exictence by its action upon objects which occupy space. When energy acts upon energy he can only look for results and try to understand the reasons.

Neither the ether theory nor the source theory can explain the transportation of all kinds of energies; therefore, a system of theorizing must be recognized which will permit light and all of the other forms of energy to be transported through the space energy field. It is possible to determine the kind of a space energy field which must exist but it is impossible to construct a mechanical model to illustrate the action taking place. The field must be of such a nature that it does not occupy space, and that it can be considered as running through and away from any point in the universe. It can be considered as running in all directions at the same instant and its velocity to be the same as the velocity of light.

The above space energy field complies with the Theory of Relativity. It cannot be mathematically explained. It can, however, be imagined. As an illustration, I release my pencil and it moves to the floor. This affects the moon, the sun, and the total universe. I cannot prove it mathematically, but I can imagine that it does. To quote Einstein, this might be one of the places where "it is more important to have imagination than it is to have knowledge."

To construct a mechanical model which will occupy space and conform to the construction of an energy field that might be considered as running past and away from a given point is beyond all probability. This cannot be done for the simple reason that one piece of metal will not pass through another piece of metal without fracturing the piece of metal. This is not true with energy because energy is constructed as a field. This field can pass through another field without interfering with the other field, providing the two fields are in synchronism. It is, however, possible to construct a mechanical model using rigid materials that will produce a motion which can be made to run past and away from any conceivable point in a given plane, such as the wabble gear mechanism.

Most text books on mechanical movements explain the action of the wabble gear. This wabble gear was used in many of the mowing machines in the early days of mechanization. The rotation of the driving wheels forced the gear to wabble but not rotate. This wabble motion was transferred as a flutter action to the cutting bar of the mowing machine. It made no difference in which direction the cutting bar was orientated. The flutter action of the cutting knives would automatically follow any new direction in which the cutter bar was operating. The cutter bar could be pointed up or down, to or from, or in any in-between direction. The flutter action of the cutting knives would, without interference, automatically follow any new direction of operation. To use laboratory parlance, the experimenter might say that energy entered the system as a sine wave by the action of the wabble gear and that the energy was expelled from the system as a flutter in any unrestricted direction at the ball and socket joint of the cutter.

It is interesting to note that a two-dimensional energy can be put in or taken out of this type of a three-dimensional energy system, by the wave motion of the wabble gear, by the pinwheel motion at the rotating stabilizer, or by the flutter type motion at the ball and socket joint of the cutter. It is also interesting to note that the height of the wave motion, the radius of the pinwheel motion, or the travel of the flutter motion can be regulated by changing the throw of the rotating stabilizer.

It is more than interesting to note that the changing of the height of the wave at the wabble gear is the same as changing the length of the wave when this wave is recorded on a plane which is at right angles to the plane of the wabble gear.

It is not only fascinating, it is intriguing to give consideration to this arrangement of a three-dimensional energy structure when studying the amplification and concentration of the two-dimensional laser energy ray.

## Chapter VI

## THE SINGING GALAXIES

A galaxy is a large aggregation of stars which are singing and dancing as they go coruscating in the space energy field. By the use of the visual telescope their ballet can be watched and recorded on film. By the use of the radio telescope their songs and melodies can be heard and recorded for study. These galaxies are truly a complete symphony in color, movement, and sound. Their pitch is accurate and distinctive, their timing is perfect and their color is informative. Their stars live and die. New stars are conceived, developed and born. Some come from dust. Some return to dust. And some do not stop at dust—they return all the way to the unseeable and the unhearable energies of space from which they originally came.

It is not the object of this dissertation to explain the theories of astronomy. It is necessary to study the theories regarding the energy action of the galaxies in order to understand the relation of the earth to the total universe. The construction of the inner earth is dependent upon the energies from outer space for its existence and state of being. It, therefore, becomes a mandate to correlate the energies of outer space with the energies of the inner earth.

The theory of relative velocities can be used to explain the actions which are taking place inside the galaxies. Likewise the theory of relative velocities can be used to explain the actions which are taking place inside the earth. Basically there is no fundamental difference between the energy action of the galaxies moving through the space energy field and the energy action of the earth moving through the space energy field. They are both dependent upon their velocity relative to the field of the universe for their existence, as bodies of material which occupy space and operate in the universe. The earth is a part of the Milky Way and the Milky Way is not at the center of the universe; neither has its position in the universe been established. For these reasons the velocity of the earth relative to the space energy field cannot be calculated relative to any known point in the universe. Neither can the speed of a galaxy be calculated from any known point.

In *The Changing Universe* John Pfeiffer explains the evolution of galaxies.* He states that "in the beginning, there is nothing but a vast irregular thing of gas clouds. The embryo galaxy has no form, only a natural tendency to spin. Form comes later. Arms appear and wrap tighter and tighter around the nucleus which becomes larger and larger." Further on he states that "a galaxy in its final stages has no spiral arms, and since it is old and may have collided with other galaxies, it has probably been swept clean of almost all its gas and dust. That means that it is sterile having lost its ability to beget new stars, which must come from interstellar matter."

Many textbooks for astronomy explain the fusion process which is taking place in the galaxies. They explain how great quantities of hydrogen atoms are in a continuous process of being transformed into helium. They also explain that great quantities of hydrogen atoms are being created in the abundance necessary to maintain the energy of the galaxies. *Astronomy* by Robert H. Baker* gives the arithmetic for the fusion process as follows:

> The relative weight of the hydrogen nucleus is 1.0076 and that of the helium nucleus is about 4.003. Where hydrogen in stellar interiors is being converted to helium, each combination of 4 hydrogen nuclei into a single helium nucleus involves a mass loss of about 4.030 - 4.003 = 0.027, or seven-tenths of one percent of the original mass. The unrecovered mass is released as energy.

* John Pfeiffer, *The Changing Universe: The Story of the New Astronomy* (London: The Scientific Book Club, undated), p. 179.

* Robert H. Baker, *Astronomy*, 7th ed. (Princeton, New Jersey: D. Van Nestrand, 1959), p. 378.

The question is, where does this released energy go? It must go traveling in the unseeable universe where it will be in a position to reenter the seeable energy and become inertial mass and thus maintain the steady state of the universe.

## Chapter VII

## WHAT FILLS SPACE?

From the time that man has been able to think and reason he has wondered how come this, how come that, and how come all things conceivable.

During the development of language and science, he has given names to many things. For the distance between the earth, the moon, other planets, and the stars, he has given the name *space*. When he speaks of the distance between two objects, he means from the center of the objects because an object does not make a void in space.

He learned that the atmosphere did not extend indefinitely out into space. He learned that sound and shock waves would not travel beyond the atmosphere. He also learned that light, heat, electricity, magnetism, and various forms of energy would pass freely beyond the atmosphere and become lost to our earth. He learned that all of these various forms of energy had a common velocity and that this velocity had a definite magnitude.

Light energy is the easiest of all the energies to observe and experiment with. It, therefore, became the first of the energies to be extensively studied in the scientific laboratories. It was determined that light, when traveling in space or vacuum, always had a constant velocity and that this velocity never changed. The experimenter learned that it is impossible to change the velocity of light by electrical means, magnetic means, mechanical means, gravitational means, or adding or subtracting light energy or any of the other various forms of energy. This constant velocity of energy has been measured to be approximately 186,000 miles per second.

For man to express his laboratory findings about light he was forced to postulate some kind of a mechanical picture to depict the action of light as it travels through prisms, lenses, the air, a vacuum, and throughout all of space. He therefore postulated the wave theory of light.

Correct or incorrect, the wave theory has, in the past, been a useful working tool for man to use when trying to comprehend the structure of the light ray, the structure of the magnetic ray, the structure of the heat ray, the structure of the electric ray, the structure of the gravitational ray, or the structure of any energy ray.

The experimenter reasoned that if wave forms of energy were to travel through space, it would be necessary to have some kind of a medium in which the wave form of light energy could propagate itself. The experimenter, therefore, invented the theory that all space must be filled with some unknown thing. He reasoned that this unknown thing must exist as a separate entity and have no connection with matter. He called this unknown thing luminiferous ether.

The luminiferous ether theory prevailed for hundreds of years. The first destructive blow to the luminiferous ether theory came when the light energy rays were proven to be discontinuous. The final and death-dealing blow to the luminiferous ether theory came when the Michelson-Morley experiments were concluded. These experiments proved without doubt that if the luminiferous ether theory was correct the world would be at rest. It was known that the world rotates; therefore, the luminiferous theory was declared incorrect and discarded.

The Michelson-Morley experiments proved that there cannot be such a thing as a rigid measuring rod. These experiments gave the experimental proof that all materials which occupy space change in length as their velocity in space is changed and that the length of any material body is determined by its velocity as it travels, in any direction, through the vast expanses of space.

Correctly or incorrectly, space is now considered by the vast majority of scientists as being filled with energy. Space is often spoken of as the space energy field. The difference in thought about the old luminiferous ether theory and the new space energy field theory is that in the ether theory the action was considered to be propagated by wave forms in a rigid medium. In the field theory the action is considered to be propagated by the interaction and reaction of fields of force. As an illustration, the field of force of a light ray is supposed to interact with the field of force in the space energy field. The field of force of the space energy field is supposed to be present inside the molecules of matter in a manner like that outside the molecules of matter. The energy throughout the space energy field may be considered as free flowing energy, without possessing mass (as we know mass) and without gravitational attraction. Free flowing energy does not possess gravitational attraction. Energy possesses gravitational attraction only after it becomes bundled and exists as a bundle of energy which occupies space. As an illustration, the pounds of energy lost to our world from the atomic explosion did not possess gravitational attraction after its bonds had been broken by the disintegrating action of the explosion. This energy became a part of the energy of the space energy field. The space energy field must have an interaction of its own to give it a finite limit and thereby maintain the steady state of the energies of the universe.

Many theories have been formulated in man's attempt to understand how energy is transported throughout the universe. The luminiferous ether theory was highly developed and then discarded. The pellet theory is known to be incorrect. It is, however, used by many mathematical scientists when it fits their needs. The wave theory has many inconsistencies but it is still used by many scientists when no other theory is adaptable to their needs.

Nearly all scientists agree that there is some kind of a field which fills all space. This field cannot be described because it cannot be captured or controlled. It cannot be seen but many of its effects can be seen, heard and measured. This field is now often spoken of as the space energy field.

This field can be considered as made up of some kind of an unexplainable something which has been named energy. According to definition, where there is energy there must be motion. This motion of the energy in the space energy field can be considered as a field of force which is running past and away from any point in the universe. This velocity can be considered as constant and equal to the velocity of light. This kind of a field cannot consist of a material structure. It must be capable of accepting any action and transport its reaction, without loss of energy, to all points of the universe. These points might be inside a body of matter or they might be where matter does not exist.

Light, electricity, magnetism, heat, gravitation, et cetera are energies which can react with the energy in the space field and be transported to all points in the universe.

Let us give thought to a very simple explanation of a man watching a railroad train passing at right angles to his line of vision and at a distance of many miles. In normal offhand speech we say that the light travels in a straight line from the train to the man and then we add that it travels at the velocity of 186,000 miles per second. If the velocity of light were instantaneous it could travel in a straight line between the train and the man. Its velocity and path of travel is, however, controlled by the characteristics of the space energy field. The controlling energy in the space energy field travels in a curve; therefore, light energy will travel in a curve as it follows the energy path of the space energy field.

A man standing on the earth watching a train moving on the earth does not realize that both he and the train are traveling through space. The velocity of the man and the velocity of the train are so slight as compared to the velocity of light that it is impossible for the man to sense that both he and the train are moving in curved lines.

Figure 1 represents a thought experiment in which the man and the train are each traveling in the same direction at one-half the speed of light and the distance from the train to the man is one-half the distance light travels in one second or 93,000 miles. The movement of the train and the observer is divided into ten equal increments. At the end of the first increment the train will have moved from A to $A_1$. The viewer will have moved from B to $B_1$ and the light ray will have moved from A to $C_1$. If the viewer and the train were stopped at $A_1$ and $B_1$ the viewer would see the train by delayed action along the dotted line from $C_1$ to $B_1$. If we repeat the above procedure to position $A_2$ and $B_2$ and $C_2$ the viewer will see the train along the dotted line from $C_2$ to $B_2$.

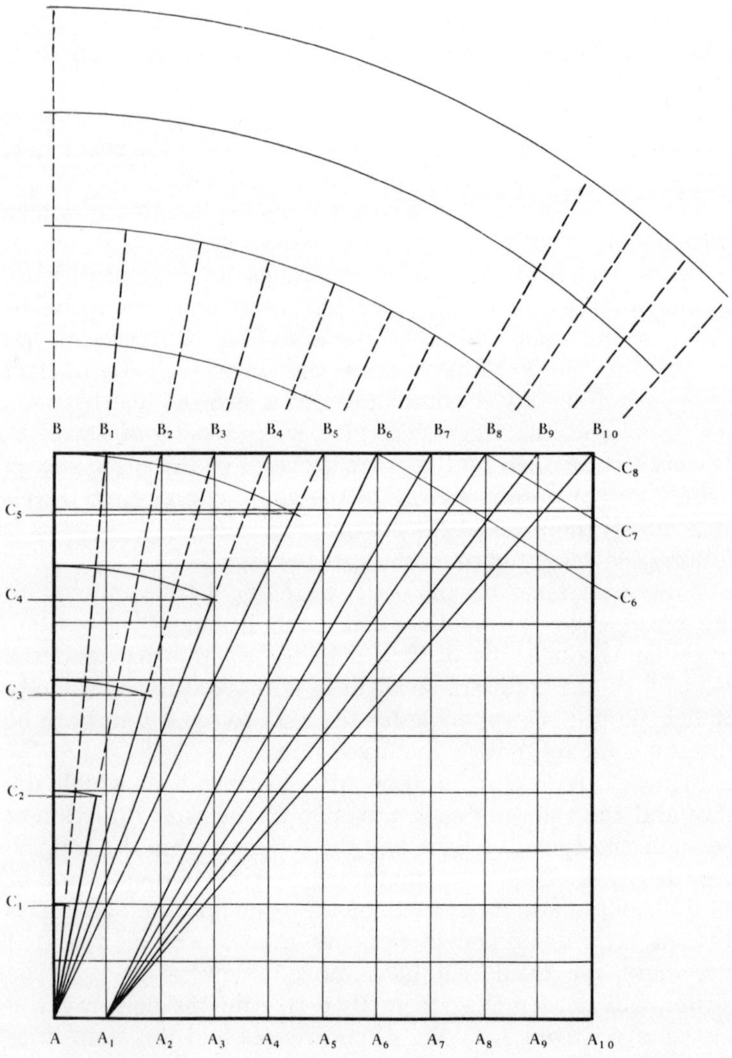
B
B1
B2
B3
B4
B5
B6
B7
B8
B9
B10
C8
C7
C6
C5
C4
C3
C2
C1
A
A1
A2
A3
A4
A5
A6
A7
A8
A9
A10

**FIGURE 1**

Likewise, for all positions from $B_2$ onward, the line of sight will be along the dotted lines from the Cs to the Bs.

Light energy can be considered as a series of discontinuous fields which travel in discontinuous jumps, each jump having no connection to any of the previous jumps. The length of each jump is a quantum of length. In Figure 1 each quantum of length was stretched out to represent two boxes on the chart. This was done for simplicity, as each box is to represent one-half the distance that light will travel in one second of time.

Let us assume that the velocity of the train and the velocity of the viewer are gradually increased until they obtain the velocity of light. It will be observed by studying Figure 1 that there will be a point along the path taken by the man when the light from the train will fail to reach the viewer. It can be concluded that the train, at this point, will have the appearance of running off into a big black hole.

When man is in a fixed position on the earth studying galaxies by the use of the telescope, he will observe that the galaxies are increasing their velocity in relation to the viewer and that the seeable part of the universe is expanding. When the velocity of the galaxy, in relation to all the other seeable objects in space, becomes equal to the velocity of light, the galaxy will have the appearance of running off into a big black hole. This can happen before the velocity of the galaxy, in relation to the viewer, becomes equal to the velocity of light. This is possible because the viewer is nearer to the center of the universe than a galaxy which is at the edge of the seeable universe, where it is unbundling itself to become a part of the energies of the unseeable universe.

One is inclined to believe that the universe has definable limits and its filled with one kind of energy which exists in two forms. These two forms blend to form the space energy fields. One form will occupy space and can be seen; the other will not occupy space and cannot be seen. The size of the seeable part of the universe is determined by its expanding velocity. The size of the unseeable part of the universe is much larger than the seeable part of the universe. Its size is determined by the slope of the energy curve of the universe.

The universe can be considered as consisting of two kinds of intermingling energies. One kind can be seen and controlled, to a limited degree, by man. The other kind cannot be seen or controlled by man. He can, however, capture and change this energy into seeable energy. It is difficult for him to comprehend, even to a limited degree, the construction of such an energy. This energy must be capable of giving the appearance of moving in all directions at one instant of time and at the velocity of light. This kind of an action cannot be explained by the use of mechanical models or visual pictures. It can be considered as moving in a field which has one more measurement than the normal dimensions of length, breadth, and thickness.

As an illustration, consider a speck of energy of infinitesimal dimensions. Move this speck in a straight line for a given distance. This will produce an energy of only one dimension, namely, length. Move this line at right angles to its length and this will produce an energy plane of two dimensions, namely, length and breadth. Move this plane at right angles to its face. This will produce an energy which will occupy space and have three dimensions, namely, length, breadth and thickness. Move each speck of energy of this three-dimension body in a direction at right angles to each face of the cube. This must be done in less than one quantum of time and at the velocity of light. This will represent an energy traveling in any conceivable direction past and away from any given point in the universe and at the velocity of light. This action cannot be made understandable by the use of mechanical models. It has, however, been explained by the creative work of mathematics and without encountering any difficulties.

## Chapter VIII

# THE STEADY STATE

The steady state theory as explained in *The Universe* by Isaac Asimov* is a theory that hydrogen is being continuously created out of nothing and is referred to as the continuous creation theory or the steady state theory.

Any theory which requires the continuous creation of something out of nothing must be based on an insecure premise. If that theory is changed to "hydrogen is continuously created out of energy," then it can be shown by the theory of relativity that the universe is at the present time capable of continuously creating hydrogen. It is not the purpose of this chapter to try to explain the origin of any energy, or an essential part, or anything concerning the total universe.

This book accepts the universe as it is now operating with its mass, matter, and energy. It accepts the theory that mass is nothing more than the appearance of energy, that mass and energy are interchangeable quantities and that the segment of the universe which is capable of being seen is finite and unbounded. This does not preclude that the space energy field can be many times larger than the visual energy field. It also accepts that part of the Einstein theory that if rays of light are dispatched in opposite directions from any point in the universe, these two rays will meet head-on at some distant point in space. And it accepts the theory that energy as mass may make its appearance at any point in the universe.

The astronomers by their system of measurements have concluded that 12.5 billion years would be the distance from the earth to a galaxy which might be receding at the velocity of light. It can be concluded that a point in space at a

* Isaac Asimov, *The Universe: From Flat Earth to Quasar* (New York: Walker and Co., 1966), p. 223.

distance of 12.5 billion light years from the earth would be the point at which all matter, if it had succeeded in traveling that far, would break down from inertial mass to become energy and pass off into the space energy field. That is, at this condition the energy in any molecule of matter becomes unbundled and returns to energy which does not occupy space, or possess inertia, or gravitational attraction, et cetera. When this energy returns to inertial mass in the steady state part of the universe it can happen to any point in the universe, even at the center of the earth.

When energy moves from the space field to the mass field it is not necessary that it enter to build a hydrogen atom and then proceed to build to a heavier atom. It can start in the iron molecule and build up to the gold molecule. It can start in the uranium 235 molecule and build up to the uranium 238 molecule.

Energy can build inertial mass at any point in the universe where the conditions are proper for its conversion. Some of the conditions for energy to convert to matter involve temperature, pressure, chemical environment, velocity, and the micro-mini inequality of the mechanical forces acting upon the atom or the molecule.

In Chapter II of this book it was shown that the earth is not at the center of the universe. For this reason it might be that light energy cannot travel in all directions to equal distances from the earth before it leaves the inertial mass field. In some directions from the earth it might be considerably less than the 12.5 billion light years.

In *Unveiling of the Universe*, Harley Woods* states:

> The rotation of the Sun may be estimated by observing the apparent movement of the sunspots. The Sun does not rotate like a solid body. At the Sun's equator, the rotation is fastest with period 25 days, at latitude 20° it is 25.7 days and at 40° 27.4 days. Other methods of determining the rotation confirm the trend of the figures and show that the rotation period is even longer nearer the Sun's poles.

* Harley Woods, *Unveiling the Universe* (New York: American Elsevier Publ. Co., 1968), p. 117.

This measurement of the rate of the sun's rotation at many locations on the sun is of the utmost importance and furnishes proof that the various parts of the sun do not rotate in unison.

*Mountains, Valleys and Subterrestrial Forces*, my first book, is fundamentally based on this peculiarity. All sections of the earth do not rotate in unison. For the sun, the difference in the rate of rotation for various sections of the ball can amount to as much as one day or more than one day in a total of ten days. For the earth, my first book considers that it is possible for various sections of the earth to have a change in velocity of rotation by as much as/or more than one day in a thousand years.

I do not believe that it is stretching the imagination beyond a reasonable limit to conclude that this peculiarity of the sun's rotation is helping to confirm my theory about the earth's creeping magma.

The astronomers, for centuries, have been measuring the distances between many of the inertial mass bodies traveling in space. These measurements show that all gaps in space which separate the heavenly bodies each from the others have been steadily increasing. As a result of these measurements, many have concluded that the total universe is expanding and is always in the process of occupying more space. This perhaps is not the proper interpretation of their measurements.

In the first part of this chapter it was shown that when an inertial mass body arrives at a distance of 12.5 billion light years from some imaginary point in the total universe its velocity will become equal to the velocity of light. When an inertial mass body (bundled energy) which occupies space travels at the velocity of light its energy will become unbundled (scattered) and it will cease to occupy space. This energy then must pass into that part of the universe which is not necessarily expanding and which must be finite and limited.

By the process of reasoning, it is to be concluded that should the free energy which fills the space between the inertial mass bodies be expanding it would then be impossible to measure the velocity of light, electricity, magnetism, heat, et cetera.

The astronomers for centuries have watched inertial mass bodies grow in outer space and they have watched other bodies decline and disappear. Some of these actions have been slow and almost impossible to detect. Others have been exceedingly rapid and explosive.

In the laboratories, the scientists have studied the interaction of energy and matter. Additional matter has been manufactured by the process of converting energy to inertial mass. And inertial mass has been destroyed by converting its controlled energy to free energy.

From the foregoing studies, it is to be concluded that it is possible to have a steady state universe which is continually converting mass to energy and energy to mass and that the inertial mass part of the universe can be continually moving outward in the free energy universe. This is true because as the particles of inertial mass move outward in the space energy field, the distance between the particles becomes greater and their relative velocities become greater. When the relative velocity of these particles becomes equal to the velocity of light they revert to free energy and join the energy of the space energy field. It is not necessary that the energy in the space energy field expand and conform to the seeable and hearable energies. It is therefore possible for the distances between the bodies in the heavens to increase without increasing the size of the total universe.

It is highly probable that a steady state universe can exist without the addition of new energy. Sufficient evidence has been gathered to build, in theory, a self-contained system in which all mass and energy will continue to undergo natural change. All inertial mass will eventually be developed from the space energy field to maintain a balance between the inertial mass field and the space energy field.

The astronomers have, by means of the visual telescope and the radio telescope, charted the life and death of inertial mass objects. Most college textbooks of astronomy explain the process by which new inertial mass is brought into being. They explain how stars are born and developed. They explain how galaxies go swirling through the space energy field and move outward from an imaginary center of the seeable part of the total universe. As the galaxies move outward the distance between the particles of mass is steadily increased. This natural increase in the distance between the particles of matter will automatically increase the velocity of each particle of matter relative to all the other particles of matter in the total universe. When the relative velocity of any inertial mass particle attains the velocity of light it will revert to free energy and take its place in the space energy field. The transfer of most of the energy from the inertial mass field to the space energy field takes place at the outer edge of the seeable universe. This does not indicate that there will be a bunching of energy at the fringe of the seeable universe because this energy is free to flow and can quickly seek its own level, like water in a pond when disturbed by a rock.

Energy from the space energy field can make its appearance as inertial mass at any point in the universe. It can originate at the center of the seeable universe or at any place where conditions will accept it. As it becomes inertial mass it will start moving toward the outer face of the universe. When it moves outward from the center it will expand and give the visual appearance of an expanding universe. The seeable universe can be considered as a small universe entwined in a larger space energy universe which can be considered as finite and limited.

There are very strong indications that the steady state universe is very active in the sun. The hydrogen content of the sun is much greater than that of the other bodies of the planets. It must be concluded that hydrogen is one of the principal products of the sun's manufacturing processes. This does not preclude that the hydrogen of the sun might be combining with energy to form new bodies of matter. To trace one of the billions of steady state cycles which are continuously in operation, we can start with the sun which is accepting energy from the space energy field and converting it to inertial mass. It is also ejecting inertial mass at a very high velocity. Some of it becomes a part of the earth. Some of it floats in space until it finds a place where it will be accepted. Some of it is swept up by the sweeping spirals of a galaxy. After it is swept up it becomes a part of the galaxy and is carried off at a speed which keeps increasing until its velocity becomes so great that it cannot be seen and it disappears into that big black hole which is described by the astronomers. In this big black hole it loses its mass and reverts to energy which has an interchangeable quality. It now becomes a part of the energy in the space energy field where it can float back to the sun and be reconverted to hydrogen et cetera. It will then have traversed one complete cycle of energy transformation in the steady state universe.

The big question now arises: What happens to humans when a planet becomes extinct and passes off into the space energy field? The answer is simple: the seeds of life can travel through the vast expanses of the universe. The astronauts brought back from outer space samples of what the scientists classed as the seeds of life. These seeds have been floating in the space energy field for billions of years. It might take several billions of years to develop humans after the seeds have taken root in an environment which is capable of supporting our kind of life.

It is to be concluded that man can destroy mass and man can destroy energy but there are no indications that man can destroy mass and energy simultaneously. In other words man can force mass and energy to interchange. Furthermore it is to be concluded that man can destroy the earth and man can destroy humans but there are no indications that man can wipe out the seeds of life.

The minimum age for our part of the universe can be determined quite accurately by studying the radioactive decay of the radioactive elements. However, the maximum age of the earth cannot be determined by this method. As an illustration, the rotation of the earth in the space energy field has caused the mountains to tumble around on the crust of the earth. During this process the old mountains have been drawn downward into the magma layer, where they have been remelted and new crystallization of the elements has taken place. The studies of radioactive decay determines when crystallization took place and has no relation to the formation of the elements of matter. The rotation of the moon in the space energy field is too slow to create sufficient heat to melt the crystallized elements. It is to be concluded, therefore, that the age of the rocks from the moon will test older than any rock which has been formed in the earth.

There are two fundamental requirements which must be fulfilled before a Steady State Universe can operate successfully.

First, there must be a condition at some point in the natural operation of the universe where the building blocks of inertial mass do not possess gravitational properties. Einstein's Theory of Relativity explains that this can happen at the instant when mass and energy are interchanging their quantities.

Second, there must be a condition where the difference in relative velocities between inertial mass and the energy in the energy field is zero. This condition for fulfillment is explained in Chapter IV.

## Chapter IX

## CRYSTALLIZATION

Crystals can be defined as a portion of matter which has a definite, orderly atomic structure.

Crystallization processes performed in the laboratories and in the industrial factories can be very well explained without referring to the Theory of Relativity. When it comes to explaining the process of crystallization within the earth it becomes necessary to include the Theory of Relativity in any explanation of the process.

As an illustration, when a large crystal of copper is grown in the laboratory, the crystal can be bent with very little force. After the crystal has been bent it will require a large force to straighten it. The reason for the large force is that during the bending operation the energy required to bend the soft copper changed the large single crystal into a mass of many tiny crystals. This process is known as work hardening. The energy required for the work hardening process comes from the energy used in the bending operation.

When crystals are formed in the innermost section of the earth, the crystals are "work hardened" by the energy derived from the expanding and contracting forces acting on the atoms. These expanding and contracting forces are the result of velocity changes brought about by the rotation of the earth in the space energy field. As the earth rotates, the space required by each atom is changed; also, the mechanical forces from outside the atom are continually changing their pressure exertion. Under the oceans these changing forces are under tension and under the mountains these changing forces are under compression. This, perhaps, is the reason that granite is never found above the magma layer in locations where the magma is under tension; it is always above the magma that is under compression.

In industry it is not advisable to coat steel wire with a

galvanized coating when the wire is to operate under tension and be subjected to vibration, as in a suspension bridge. The pumping action of the vibration will cause some of the atoms of the galvanizing mixture to disassociate themselves from the galvanizing structure and become associated with the molecules of the steel structure. This new arrangement of the molecules will downgrade the tensile strength of the molecular arrangement of the crystallized steel and destroy the strength of the steel cable.

When large diamond crystals are formed in the earth there are cavities left in many of the crystal formations. Many of these voids are filled after the original process of crystallization has been completed. In finished diamonds the voids in the crystalline structure which have not been filled are known as imperfections. To force quantums of carbon energy to work their way into a crystal to fill an imperfection there must be some kind of a pumping action whereby the forces exerted by the earth are continually increasing and decreasing. This pumping action can work an atom through a crystal. The mechanics of the action is similar to that of the galvanizing material on the steel cable of the suspension bridge. Due to the stress and strain relation of the bundled energies there must be times when their velocity, relative each to the other, is zero. At this condition the mechanical force gradient will determine the direction of travel.

*Elements of Structural Geology* by E. Sherbon Hills F R S* discusses salt domes and states that "students of the Roumanian salt plugs have naturally suggested that orogenic forces were mainly responsible for the salt injections, but some other explanation is clearly required where beds are flat-lying." If we accept the Theory of Relativity and its relation to orogeny, it can be explained by this theory why some of the crystallized salt beds are flat-lying and others are vertical salt domes. As an explanation, the creeping magma is forcing mountains up and pulling trenches down. During this tumbling of the crust of the earth, mountains are tumbled

* E. Sherbon Hills, *Elements of Structural Geology* (New York: John Wiley and Sons, 1963), p. 270.

into salt deposits. The salt deposits with their absorbed moisture and gases are drawn down into the magma layer. In the magma layer the absorbed moisture and gases are driven off by the heat of the magma. Due to the high temperature of the magma the moisture and the gases are maintained in the form of bubbles. The salt and bubbles will creep along with the magma layer because they are a part of the magma layer. As the magma creeps forward beneath the crust of the earth the outer portion of the magma layer decreases in temperature. When temperature and other conditions are proper for crystallization the salt will crystallize. If crystallization takes place where the magma is under tension the new salt beds will be flat-lying. If crystallization takes place where the magma layer is under compression the salt domes will be vertical.

The salt beds and salt domes will follow closely with the granite processes of a mountain building. The Devil's Watchtower in Wyoming is an excellent example of vertical crystallization. It is to be noted, however, that salt domes will absorb more moisture and gases than granite domes will absorb. As a salt dome is pushed up it becomes cooled; the vapor and the gases are reabsorbed. This will increase the specific gravity of the mixture and the salt dome will become stable beneath the crust of the earth. If the salt dome had not cooled and absorbed the moisture and the gases, the dome would have continued to rise. At the surface the vapors and the gases would have been expelled as a volcano.

From the foregoing it can be seen why salt domes do not pierce the outer crust of the earth but remain in the neighborhood of 1,000 feet below the surface. Here the temperature gradient is the major controlling factor because it is the primary influence for the molecular arrangement of the mixture which determines the specific gravity of the salt with its moisture and gas content.

In some cases of mountain building the cooling of the magma has not taken place until after a portion of the vapors and gases has been expelled from the magma. The expansion of the vapors and gases has produced a refrigeration effect which has cooled the magma. The cooled magma has crystallized and reabsorbed the bubbles of moisture and gases. This action has increased the specific gravity of the magma and as a result the magma has contracted and formed a void. The crust has fallen into the void and formed a crater such as Crater Lake, Oregon.

## Chapter X

## GRAVITATION

Albert Einstein showed that gravitation exists as a consequence of inertial force acting on bodies in curved space. This is known as the Einstein equivalence principle.

As we try to understand the above statement let us consider the development of the theory of gravitation from the time that Sir Isaac Newton was struck by a falling apple. This caused him to have serious thoughts about gravity. After much study and many experiments he came to the conclusion that every particle of matter attracts every other particle of matter directly as their mass and inversely as the square of the distance between the masses. This law of gravitation is still as valid in these days of Einsteinian mechanics as it was in the day of Newtonian mechanics. The only difference is that mass is not a constant; i.e., on the earth a one-pound weight of iron at a specific instant of time does not weigh one pound at the next instant of time. Also, the length of the measuring rod is undergoing change and the timing clock is continually speeding up or slowing down. To apply Newton's laws of gravity to their fullest extent, it will be necessary to add many mathematical arrangements, whereby these small changes in mass, distance, and clock speeds can be taken into consideration.

The space energy field is operating inside the earth in a manner like that in which it is operating in the space outside of the earth. Let us posit an illustration. When the apple fell to the ground it came to rest relative to the surface of the earth. Let us assume that the apple fell into a large hole in the ground and that this hole extended to the center of the earth and beyond. If the velocity in space is not taken into consideration (Newtonian mechanics), the theory is that the apple would come to rest, suspended by gravity, at a point in the center of the earth. If the velocity in space is taken into consideration (Einsteinian mechanics), the apple would, in theory, not come to rest at the center of the earth. It would be suspended by gravity and remain approximately midway between the two poles, but not on the axis of the earth's rotation. It would remain expelled at some distance from the earth's axis and it would not rotate with the earth. If all sections of the earth—i.e., the crust of the earth, the magma layer of the earth, and the center core of the earth—were rotating in unison the apple would complete one revolution of gyration per each revolution of the crust of the earth. This is true because the apple is at rest in the space gravitational field and the earth is rotating. The number of revolutions of the crust of the earth does not coincide with the number of revolutions of the core of the earth. It will therefore be impossible to determine the exact speed of gyration of the apple near the center of the earth, i.e., relative to the crust of the earth. It is estimated that it might increase as much as 360 degrees in 1,000 years, i.e., over the one revolution per day of the crust of the earth.

To explain why the apple will gyrate off the center of the earth it will be necessary to take into consideration that the mass of every particle of matter will be determined by its velocity relative to the space energy field. The rotation of the earth is continually changing the velocity of each particle of matter as it goes swirling in space. Since the weight (gravitational attraction) of each particle of matter is changed each instant that its velocity is changed, one directional side of the earth will always be heavier than the opposite directional side of the earth. Therefore the center of gravitational attraction will never coincide with the center of the ball. This shows that to calculate correctly the gravitational attraction between the earth and all other bodies of matter it will be necessary to devise some kind of a formula which will take into account the shifting of the center of gravitation of the earth and the shifting of the center of gravitation of all bodies of matter. If the foregoing could be accomplished and added to Newtonian mechanics, then Newtonian mechanics would be sufficient to make the necessary calculations. This is not possible with our present knowledge. It is only possible to do as Einstein did. He bunched everything into two simple formulas. Man is now trying to understand these formulas.

Gravitational rays have never been captured. Some scientists believe that they have been detected. If gravitational rays exist, they are probably similar to electric rays in the respect that the direction of the traveling force cannot be determined, i.e., it will be impossible to determine if the force is coming or going. When a man is sitting on the earth it appears to be more understandable to say that gravitational attraction pulls him toward the center of the earth than to say that the gravity pushes him toward the center of the earth. Consider a man sitting on the apple which is forced in a circular movement relative to the center of the earth. It becomes more natural to say that he is pushed around at the center of the earth than it is to say that he is pulled around at the center of the earth.

The Foucault pendulum experiments were performed to show that the earth was rotating in the space energy field. The Foucault pendulum is a heavy weight suspended by a long wire from a fixed point in a high tower. Because the wire is long and the bob is heavy the oscillations will be slow and will continue for months after it has received the inertial energy to start the bob oscillating. If the pendulum is located at the North Pole or the South Pole, the earth will rotate below it and the pendulum will continue to oscillate in the same plane, relative to the space energy field, in which it was started to oscillate. The rotation of the earth will not affect the direction of oscillation of the bob of the pendulum. Any action to change the direction of oscillation must come from outside the earth, i.e., from the gravitational field of the universe.

In the previous illustration the apple was standing still relative to the gravitational field and the earth was rotating around it. In that illustration the force which was holding the apple from rotating with the earth was coming not from outside the earth but from inside the earth. The force from outside the earth was acting upon the individual particles of the earth to change their gravitational force and the gravitational force was forcing the apple to remain on the heavier side of the earth, i.e., on the side of the earth where its velocity in relation to the space energy field was the greater. This heavy side did not rotate with the earth but remained stationary relative to the space energy field. It was the gravitational forces from the heavy side of the earth which forced the apple from the axis of the earth and stopped the apple from following the inertial masses which were rotating about the axis of the rotating earth.

The gyration of the North Pole and the South Pole is such that it can be measured relative to the crust of the earth. These measurements tend to substantiate the foregoing line of reasoning.

## CONCLUSIONS

First: The earth, for its existence and state of being, is dependent on the energies from distances far beyond our galaxy.

Second: The earth is a part of the steady state of the total universe.

## HIGHLIGHTS

37 The size of of the seeable part of the universe is determined by velocity.
38 An energy with more than three dimensions.
39 The steady state.
39 A galaxy receding at the velocity of light.
41 The sun does not rotate as a ball.
41 The free energy field not expanding.
42 New inertial mass brought into being.
43 The space energy universe can be considered as finite and limited.
43 The steady state is active in the sun.
45 The age of rocks on the moon will test older than the rocks on the earth.
46 Crystallization.
46 Crystals are work hardened.
47 The pumping action can work an atom through a crystal.
48 Salt beds and salt plugs.
50 Gravitation is an inertial force acting on bodies in curved space.
51 The center of gravity of the ball does not rotate with the earth.
52 The shifting of the center of the gravity of the earth.
52 The Foucault pendulum.
53 The gravitational force from the heavy side of the earth.
53 The gyration of the poles of the earth.